AF555964

YLP®

Young Learner's

MENTAL MATHS

Book-2

Rajesh Singh

1 2 3 4 5 6 7 8 9 10

© Young Learner Publications®, India

CONTENTS

Published by:

YOUNG LEARNER PUBLICATIONS®

G-1A Rattan Jyoti, 18 Rajendra Place
New Delhi-110008 (INDIA)
Tel.: 011-25750801, 25820556, 25755559
Fax: 91-11-25764396
Email: goodwillpub@gmail.com
gph.ylp@gmail.com
gph.ylp@goodwillpublishinghouse.com
Website: www.goodwillpublishinghouse.com

© Young Learner Publications, India

All rights reserved. No part of this publication may be reproduced, stored in a retrieval system or transmitted, in any form or by any means—mechanical, photocopying, recording or otherwise, without prior written permission of Young Learner Publications, India.

Worksheet - 1

Counting

1. Count forward and fill in the boxes.

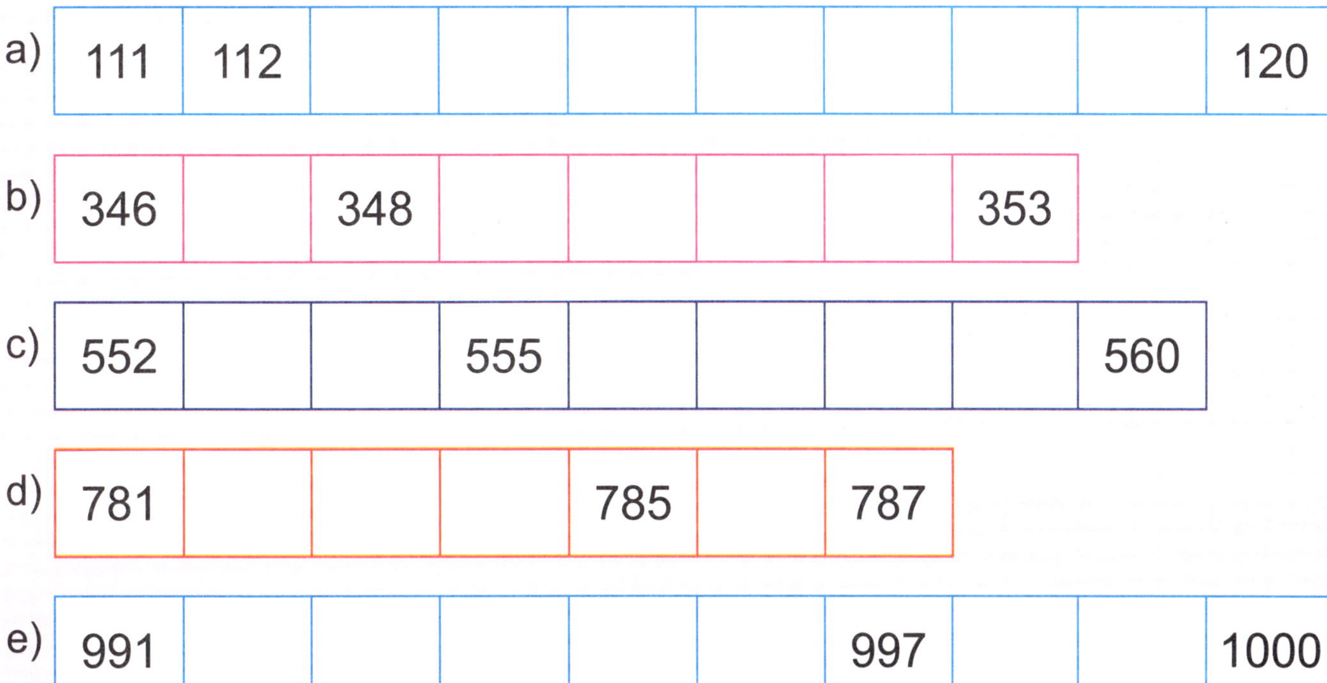

a)	111	112								120
b)	346		348					353		
c)	552			555					560	
d)	781				785		787			
e)	991						997			1000

2. Count backward and fill in the boxes.

a)	260							253	
b)	433			430					425
c)	615				611		609		
d)	702							695	
e)	876						870		

Worksheet - 2

Number Names and Numerals

Number name	Numeral
One hundred twenty-six	
Two hundred forty-five	
Three hundred sixty-nine	
Four hundred thirty-four	
Four hundred ninety-six	
Five hundred twelve	
Five hundred sixty-three	
Six hundred fifty-one	
Six hundred eighty	
Seven hundred forty	
Seven hundred eighty	
Eight hundred forty-three	
Eight hundred ninety	
Nine hundred thirty	
Nine hundred ninety-two	

Numeral	Number name
185	
218	
377	
462	
489	
529	
580	
643	
671	
732	
772	
840	
875	
929	
999	

Worksheet - 3

Ones and Tens
Before, After and In Between

1. Fill in the blanks.

a) 347 ______ hundreds ______ tens ______ ones

b) Four hundred twenty-eight ______ hundreds ______ tens ______ ones

2. Write the numeral for:

a) 5 hundreds 6 tens 2 ones ________

b) 6 hundreds 5 tens 9 ones ________

3. Write the number name for:

a) 7 hundreds 7 tens 4 ones ______________________________

b) 8 hundreds 1 ten 6 ones ______________________________

4. What comes before?

a) ______, 437

b) ______, 655

c) ______, ______, 952

d) ______, ______, ______, 382

5. What comes after?

a) 235, ______

b) 531, ______

c) 873, ______, ______

d) 462, ______, ______, ______

6. What comes in between?

a) 223, ______, 225

b) 325, ______, 327

c) 705, ______, ______, 708

d) 996, ______, ______, ______, 1000

7. What comes before and after?

a) ______, 170, ______

b) ______, ______, 562, ______, ______

c) ______, ______, ______, 997, ______, ______, ______

Worksheet - 4

Place Value of Digits, Expanded Form, Comparing Numbers Ascending and Descending Order

1. Write digit in hundreds, tens and ones place.

a) 259: Digit in H's place ______, digit in T's place ______, digit in O's place ______.

b) 638: Digit in H's place ______, digit in T's place ______, digit in O's place ______.

2. Write the Place Values (PV).

a) 417: PV of 4 is ________, PV of 1 is ________ and PV of 7 is ________.

b) 780: PV of 7 is ________, PV of 8 is ________ and PV of 0 is ________.

3. Write the expanded form.

a) 365 = _____ + _____ + _____   b) 998 = _____ + _____ + _____

4. Fill in the blanks.

a) The numeral with 3 at hundreds place, 2 at tens place and 4 at ones place is ________.

b) The numeral with place value of 5 as 500, 1 as 10 and 8 as 8 is ________.

c) The numeral with expanded form 800 + 30 + 6 is ________.

5. Put the correct sign (< or = or >) in the blanks.

a)	458___378	b)	527___527	c)	217___675
d)	629___629	e)	972___885	f)	999___1000

6. Write in descending order.

a)	534, 356, 952, 348	
b)	414, 567, 449, 834, 352	

7. Write in ascending order.

a)	354, 351, 753, 667	
b)	922, 553, 314, 467, 500	

Worksheet - 5

Smallest and Greatest Numbers with Given Digits
Rounding-off Numbers

1. **Make the greatest and smallest 3 digit numbers using 0, 2 and 4. Repetition is allowed.**

   Greatest: ________________ Smallest: ________________

2. **Make the greatest and smallest 3 digit numbers using 1, 3, 5 and 7. Repetition is not allowed.**

   Greatest: ________________ Smallest: ________________

3. **Write all the possible 3 digit numbers greater than 500 using the digits 0, 5 and 6. Repetition is not allowed.**

   ____

   Greatest: ________________ Smallest: ________________

4. **Round the following to the nearest ten.**

   a) 131 ≈ ________________ b) 234 ≈ ________________

   c) 568 ≈ ________________ d) 755 ≈ ________________

5. **A bag has 834 brushes. Number of brushes rounded to nearest ten is ________________.**

6. **Round the following to the nearest hundred.**

   a) 344 ≈ ________________ b) 459 ≈ ________________

   c) 891 ≈ ________________ d) 987 ≈ ________________

7. **An exhibition was visited by 592 people today. Number of visitors rounded to nearest hundred is ______________.**

8. **Put the correct sign (< or >).**

   706 (rounded to nearest ten) _____ 687 (rounded to nearest hundred)

Note: ≈ denotes rounding off.

Worksheet - 6

Addition Without Carrying and With Carrying

Solve the following:

1.

5	1	5
+ 3	0	2

2.

5	2	1
+ 1	4	6

3.

5	8	0
+ 2	1	4

4.

6	2	1
+ 1	0	7

5.

7	3	4
+ 1	8	2

6.

4	4	7
+ 3	7	2

7.

1	9	8
+ 3	6	0

8.

	8	3
+	4	2

9.

2	3	4
+ 1	8	7

10.

3	4	7
+ 3	7	5

11.

2	9	8
+ 1	6	3

12.

	8	6
+	4	9

13. 132 + 245 = ____________
14. 231 + 548 = ____________
15. 341 + 414 = ____________
16. 353 + 546 = ____________
17. 238 + 147 = ____________
18. 254 + 351 = ____________
19. 345 + 276 = ____________
20. 345 + 654 = ____________

Worksheet - 7

Adding Three Numbers

Solve the following:

1

3	1	0
1	4	5
+ 2	2	2

2

2	1	6
1	2	1
+ 4	2	2

3

1	2	3
3	4	6
+ 1	0	1

4

5	2	0
1	1	2
+ 3	5	5

5

	9	4
		6
+		8

6

	8	3
	9	1
+		5

7

	8	3
	9	3
+	3	3

8

	3	6
	2	1
+	9	7

9

3	0	7
1	3	3
+	2	1

10

2	1	2
4	2	1
+	3	8

11

2	8	5
1	9	3
+	8	0

12

4	5	6
1	8	1
+	2	7

13

3	7	4
1	2	1
+ 1	2	2

14

4	8	1
1	2	1
+ 2	2	0

15

2	0	1
2	3	4
+ 1	9	8

16

2	7	9
1	6	9
+ 4	8	5

17 123 + 105 + 210 = ____________

18 201 + 120 + 312 = ____________

19 134 + 125 + 421 = ____________

20 235 + 132 + 63 = ____________

Worksheet - 8

Addition Stories

1. A gym had 247 members. 123 people more joined it. How many members does the gym have now?

2. A mobile store sold 265 mobile phones in the first month and 348 mobile phones in the second month. How many phones were sold in two months?

3. In a one-day international cricket match, India scored 325 runs and Australia scored 298 runs. How many total runs were scored in the match?

4. A factory manufactured 143 items on the first day, 289 items on the second day and 406 items on the third day. How many total items were manufactured in three days?

5. There are three bags. First bag contains 234 brushes, second bag has 102 brushes, third bag has 165 brushes. How many brushes are there in the three bags?

Worksheet - 9

Estimating Sum and Addition Facts

1. Estimate the sum by rounding to nearest ten.

a) 226 + 159 ≈ __________ + __________ = __________

b) 463 + 315 ≈ __________ + __________ = __________

c) 231 + 452 ≈ __________ + __________ = __________

2. A packet has 168 buttons. Another packet has 132 buttons. Estimate the total number of buttons in the two packets by rounding to the nearest ten.

__________ + __________ ≈ __________ + __________ = __________

3. Estimate the sum by rounding to the nearest hundred.

a) 431 + 249 ≈ __________ + __________ = __________

b) 351 + 497 ≈ __________ + __________ = __________

c) 171 + 627 ≈ __________ + __________ = __________

4. A bag has 425 combs. Another bag has 377 combs. Estimate the total number of combs in the two bags by rounding to the nearest hundred.

__________ + __________ ≈ __________ + __________ = __________

5. Fill in the blanks.

a) 519 + ______ = 242 + 519

b) ______ + 319 = 319 + 423

c) 222 + 438 = 438 + ______

d) 280 + 526 = ______ + 280

e) 532 + 0 = ______

f) 247 + ______ = 247

g) ______ + 0 = 614

h) 0 + 725 = ______

i) 0 + ______ = 829

j) ______ + 521 = 521

Worksheet - 10

Subtraction Without Borrowing and With Borrowing

Solve the following:

1.

6	8	9
– 4	0	6

2.

7	2	0
– 3	2	0

3.

8	9	6
– 5	3	2

4.

9	0	8
– 2	0	5

5.

9	6	5
– 3	3	7

6.

8	8	0
– 2	2	5

7.

9	2	7
– 4	7	3

8.

9	0	8
– 6	2	6

9.

7	2	4
– 3	3	6

10.

8	3	0
– 5	5	6

11.

7	0	1
– 4	9	9

12.

9	0	6
– 6	7	8

13. 775 – 245 = ____________
14. 689 – 286 = ____________
15. 558 – 317 = ____________
16. 775 – 475 = ____________
17. 971 – 547 = ____________
18. 825 – 582 = ____________
19. 824 – 386 = ____________
20. 905 – 672 = ____________

Worksheet - 11

Addition and Subtraction Together

Solve the following:

1. 667 – 432 + 246

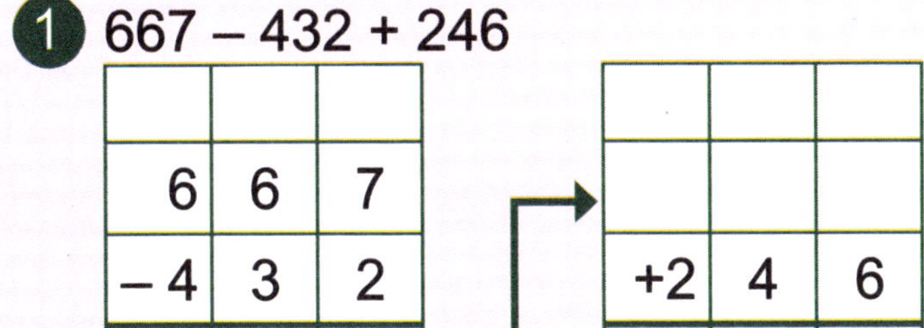

2. 593 + 344 – 297

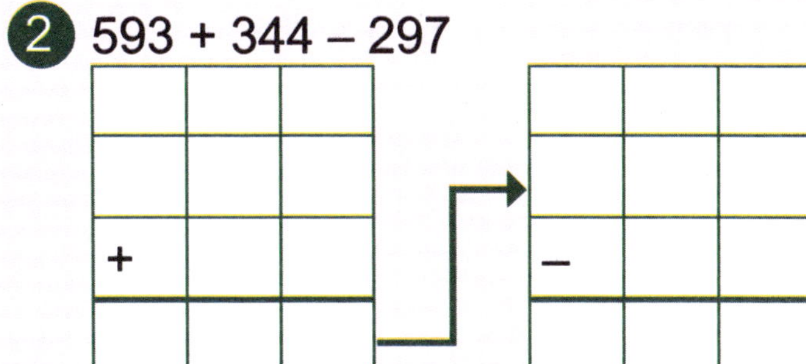

3. 878 – 613 + 129

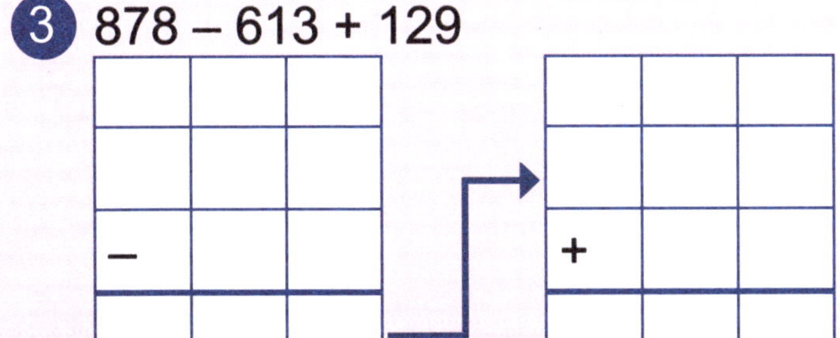

4. 243 + 459 – 292

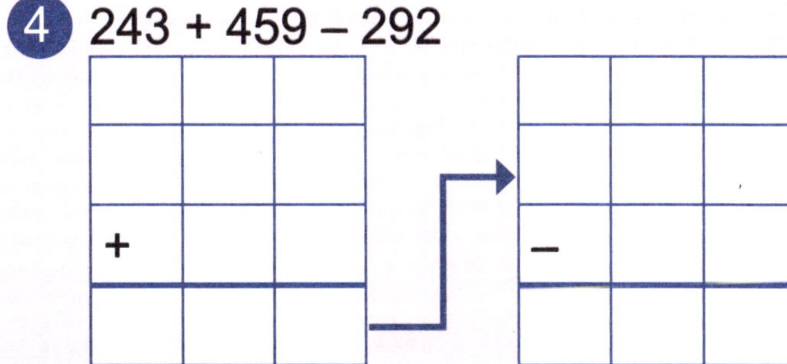

5. 464 + 326 – 135

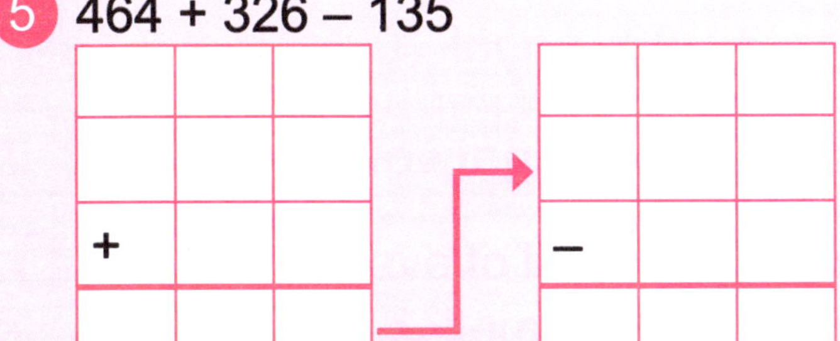

6. 662 – 553 + 373

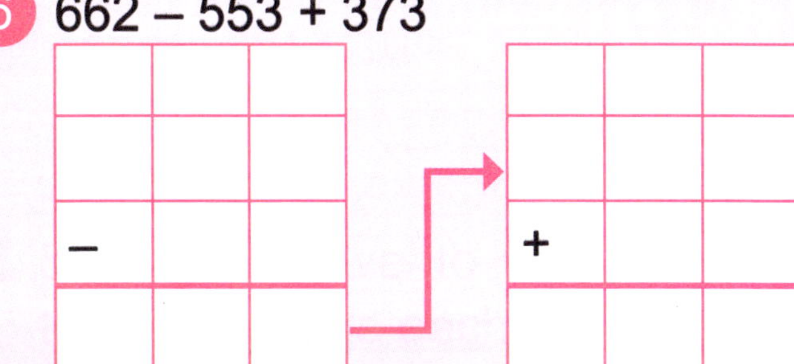

7. 805 – 416 + 302

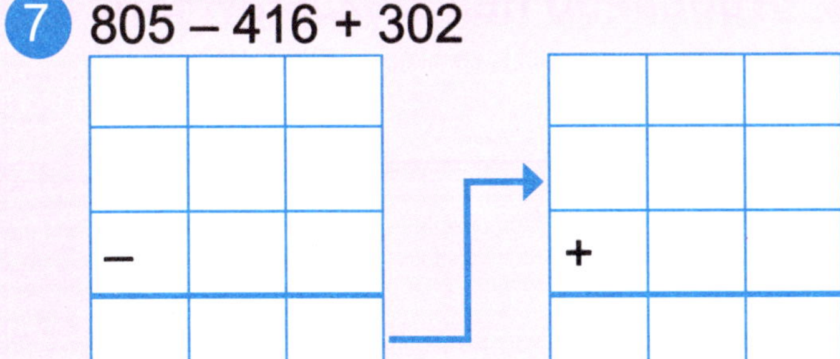

8. 366 – 243 + 232

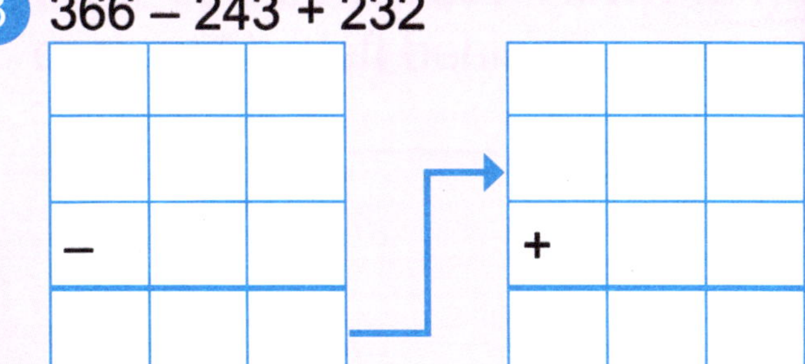

9. 554 + 247 – 451

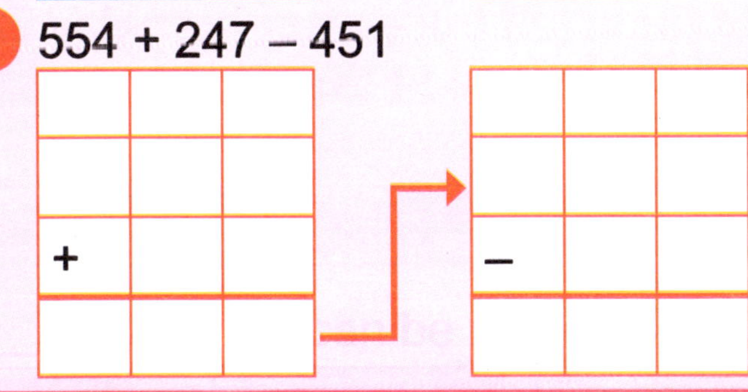

10. 700 + 200 – 429

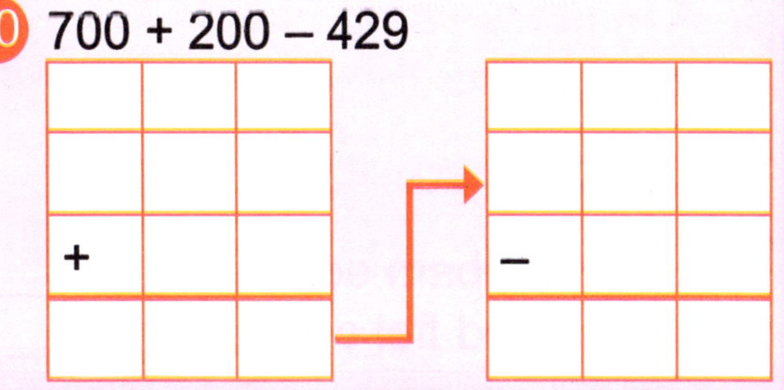

Worksheet - 12

Subtraction Stories

1. A bookseller had 698 copies of a book. He sold 235 of them. How many books are left with him now?

2. Lara had 280 novels. She donated 152 of them. How many novels are left with her now?

3. A fruit seller had 520 apples. He sold 469 of them. How many apples are left with him now?

4. There are 425 passengers travelling in a train. At a station 176 get off and 212 new passengers board the train. How many passengers are there in the train now?

5. A military camp had 913 men. 528 of them left the camp. Next day 220 men joined the camp. How many men are there in the camp now?

Worksheet - 13

Estimating Difference and Subtraction Facts

1. Estimate the difference by rounding to the nearest ten.

a) 898 – 129 ≈ __________ – __________ = __________

b) 964 – 615 ≈ __________ – __________ = __________

c) 762 – 513 ≈ __________ – __________ = __________

2. Packet A has 775 toffees. Packet B has 359 toffees. Estimate how many more toffees does packet A have by rounding to nearest tens.

__________ – __________ ≈ __________ – __________ = __________

3. Estimate the sum by rounding to the nearest hundred.

a) 764 – 483 ≈ __________ – __________ = __________

b) 915 – 522 ≈ __________ – __________ = __________

c) 718 – 235 ≈ __________ – __________ = __________

4. The first bag has 568 balls. The second bag has 677 balls. Estimate how many balls less does the first bag have by rounding to nearest hundreds.

__________ – __________ ≈ __________ – __________ = __________

5. Fill in the blanks.

a) _____ – 672 = 0

b) 739 – _____ = 0

c) 369 – 369 = _____

d) _____ – 0 = 756

e) 845 – _____ = 845

f) 533 – 0 = _____

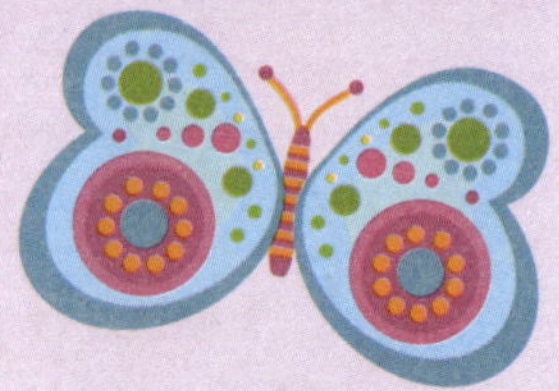

Worksheet - 14

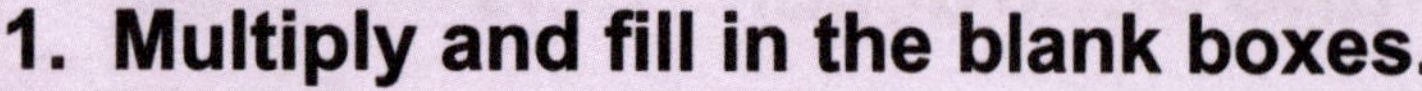

Tables 6 to 10

1. **Multiply and fill in the blank boxes.**

×	6	7	8	9	10
1					
2					
3					
4					
5					
6					
7					
8					
9					
10					

2. **Each bundle has 9 notebooks. How many notebooks are there in 6 bundles?**

   Number of notebooks = ______ × ______ = ______

3. **Each box has 4 brushes. How many brushes are there in 7 boxes?**

   Number of brushes = ______ × ______ = ______

4. **Each pouch has 7 erasers. How many erasers are there in 8 pouches?**

   Number of erasers = ______ × ______ = ______

5. **Each team has 5 men. How many men are there in 9 teams?**

   Number of men = ______ × ______ = ______

6. **Each boat can carry 5 passengers. How many passengers can 10 boats carry?**

   Number of passengers = ______ × ______ = ______

Worksheet - 15

Multiplication Without Carrying and With Carrying

Solve the following:

1.

1	0	1
×		6

2.

1	1	0
×		7

3.

1	1	1
×		8

4.

1	0	2
×		9

5.

1	4	3
×		6

6.

1	5	8
×		6

7.

1	2	5
×		7

8.

1	4	2
×		7

9.

1	1	7
×		8

10.

1	2	3
×		8

11.

1	0	5
×		9

12.

1	0	2
×		9

13. 110 × 6 = ____
14. 101 × 7 = ____
15. 100 × 8 = ____
16. 101 × 9 = ____
17. 157 × 6 = ____
18. 127 × 7 = ____
19. 119 × 8 = ____
20. 107 × 9 = ____
21. 76 × 10 = ____
22. 45 × 10 = ____
23. 72 × 10 = ____
24. 87 × 10 = ____

Hint: Whenever you have to multiply a number by 10, simply put '0' after the number to get the answer.

Worksheet - 16

Multiplication Stories

1. A packet has 142 biscuits. How many biscuits will 6 such packets have?

2. Each box has 126 hairpins. How many hairpins will 7 such boxes have?

3. A ferry can carry 115 passengers in a trip. How many passengers will it carry in 8 trips?

4. Each pouch has 106 buttons. How many buttons will 9 such pouches have?

5. Each bundle has 96 sheets. How many sheets will 10 such bundles have?

6. A box has 24 chocolates in it. How many chocolates will 7 such boxes have?

7. A carton has 27 bottles. How many bottles will 6 such cartons have?

Worksheet - 17

Multiplication Facts

1. Fill in the blanks.

a) A × B = _______ × A

b) [A × B] × C = A × [_______ × C]

c) A × _______ = A = 1 × _______

d) A × _______ = 0 = 0 × _______

2. Solve the following:

a) 324 × _______ = 209 × 324

b) _______ × 113 = 113 × 510

c) 512 × 219 = 219 × _______

d) 563 × 172 = _______ × 563

e) 239 × _______ = 623 × _______

f) _______ × 731 = _______ × 882

g) [139 × 626] × 563 = 139 × [626 × _______]

h) [528 × 242] × 996 = 528 × [_______ × 996]

i) [372 × 868] × 263 = _________ × [868 × 263]

j) [469 × 425] × _________ = 469 × [425 × 673]

k) [352 × _______] × 466 = 352 × [526 × 466]

l) [_______ × 342] × 567 = 598 × [342 × 567]

m) 261 × 1 = _______

n) _______ × 1 = 519

o) 139 × _______ = 139

p) 1 × 611 = _______

q) 1 × _______ = 872

r) _______ × 925 = 925

s) 496 × 0 = _______

t) 738 × _______ = 0

u) 0 × 634 = _______

v) _______ × 419 = 0

w) 321 × 727 × 0 = _______

x) 762 × 0 × 877 = _______

y) 0 × 16 × 427 = _______

z) 193 × 82 × 1 × 0 = _______

Worksheet - 18

Long Division Without Remainder and With Remainder

Solve the following:

1 $6\overline{)366}$ Quotient is ______.	2 $7\overline{)651}$ Quotient is ______.	3 $8\overline{)936}$ Quotient is ______.
4 $9\overline{)324}$ Quotient is ______.	5 $10\overline{)820}$ Quotient is ______.	6 $6\overline{)458}$ Quotient is ______. Remainder is ______.
7 $7\overline{)347}$ Quotient is ______. Remainder is ______.	8 $8\overline{)535}$ Quotient is ______. Remainder is ______.	9 $9\overline{)715}$ Quotient is ______. Remainder is ______.

Worksheet - 19

Division Stories and Division Facts

1. Solve the following:

a) 413 ÷ 413 = ______ b) 598 ÷ ______ = 1 c) ______ ÷ 193 = 1

d) 822 ÷ 822 = ______ e) ______ ÷ 1 = 841 f) 813 ÷ ______ = 813

2. Fill in the blanks.

a) 10 × 6 = 60 then 60 ÷ 10 = _____ and 60 ÷ 6 = _____

b) 5 × 7 = 35 then 35 ÷ 5 = _____ and 35 ÷ 7 = _____

c) 5 × 8 = 40 then 40 ÷ 5 = _____ and 40 ÷ 8 = _____

d) 4 × 9 = 36 then 36 ÷ 4 = _____ and 36 ÷ 9 = _____

e) 6 × 10 = 60 then 60 ÷ 6 = _____ and 60 ÷ 10 = _____

3. Solve the following:

a) 847 items have to be packed into packets of 7 items each. How many packets can be made?

______ packets can be made.

b) 588 eggs have to packed into trays of 8 eggs each. How many trays can be made? How many eggs will be left behind?

______ trays can be made.
______ eggs will be left behind.

Worksheet - 20

Measuring Length with Non-Standard Units

Fill in the blanks.

1. The table is _____ handspans long.

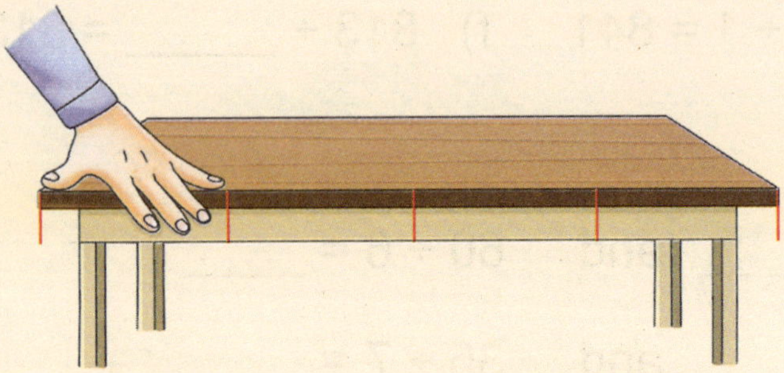

2. The carpet is _____ paces long.

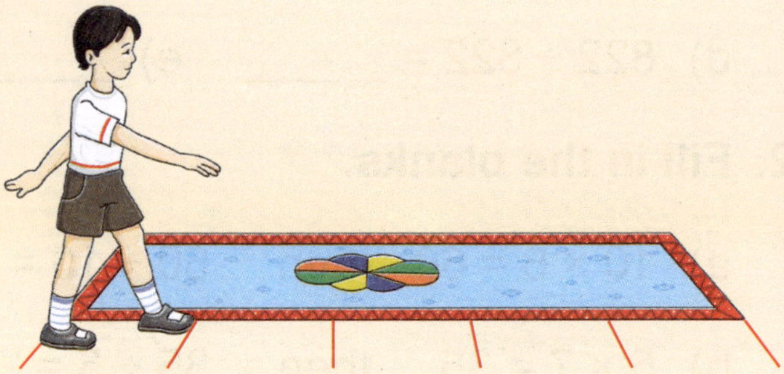

3. The window is _____ cubits long.

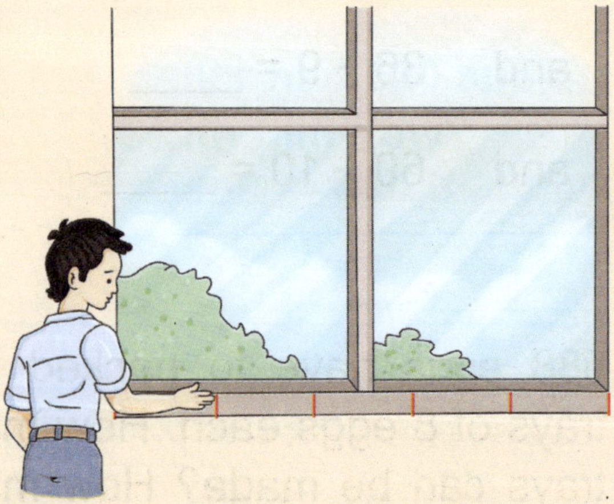

4. The bat is _____ pencils long.

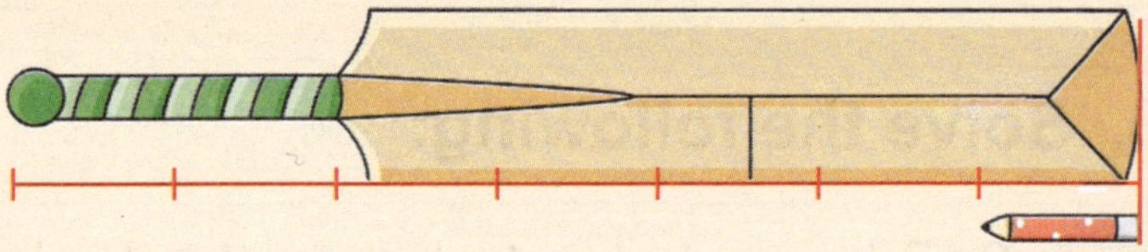

5. The tree is ____ paces from the pole.

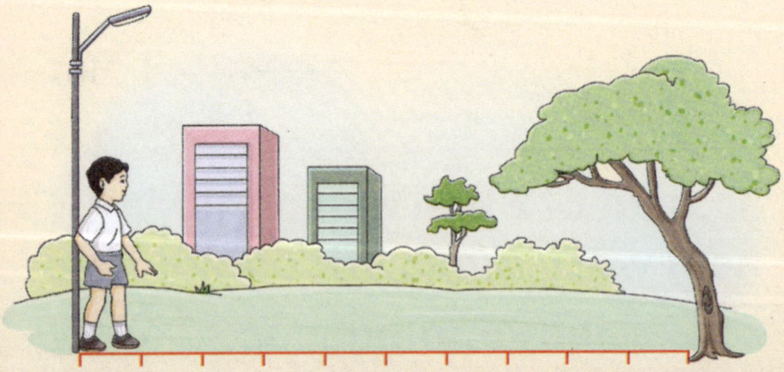

6. The rope is _____ cubits long.

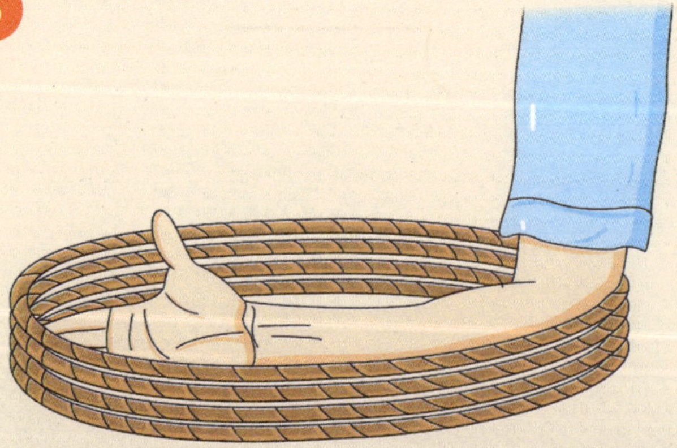

Worksheet - 21

Scale, Broken Scale and Choosing the Correct Unit

1. Find the length using the given scale.

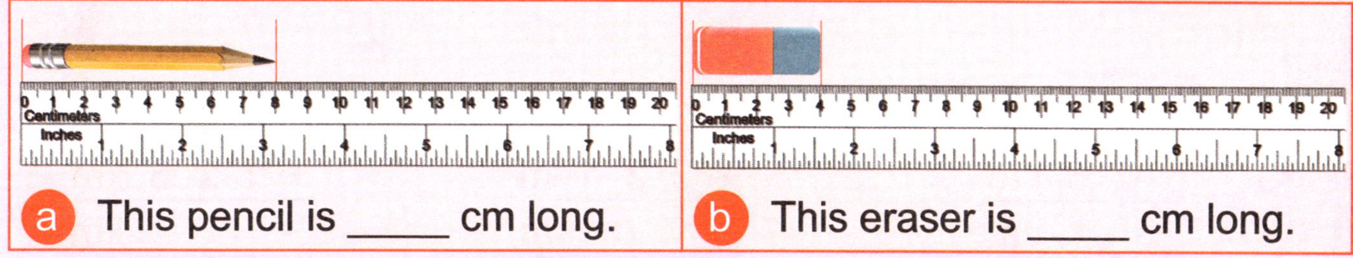

a) This pencil is _____ cm long.

b) This eraser is _____ cm long.

2. Find the length using the broken scale.

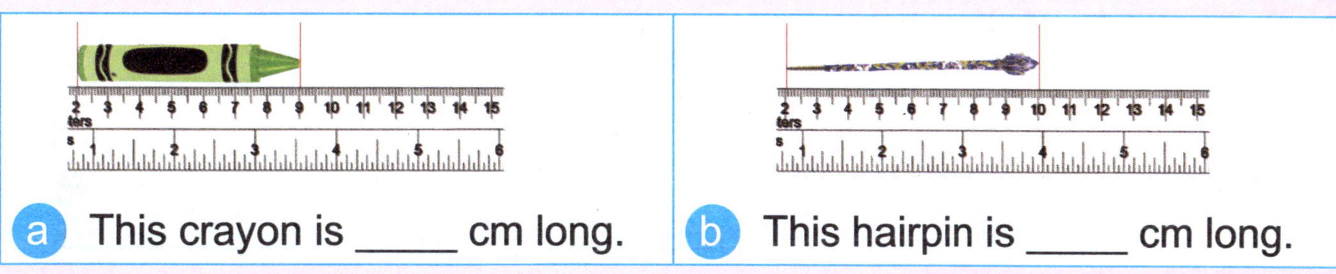

a) This crayon is _____ cm long.

b) This hairpin is _____ cm long.

3. Fill in the blanks.

a) Units used to measure length are centimetre (cm), ______________________ ____.

b) 1 m = ______________ cm

c) 1 km = ______________ m

4. What unit will you use to measure the following? Tick the correct choice.

a)	Distance between Delhi and Mumbai	cm	m	km
b)	Height of a mobile tower	cm	m	km
c)	Side of a drawing sheet	cm	m	km
d)	Length of a highway	cm	m	km
e)	Length of a pencil	cm	m	km
f)	Length of a pair of scissors	cm	m	km
g)	Length of a spoon	cm	m	km
h)	Height of an electric pole	cm	m	km

Worksheet - 22

Addition, Subtraction, Multiplication, Division of Lengths

1. Add the following:

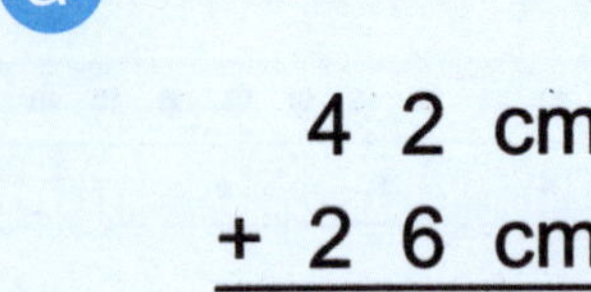

a	b	c
4 2 cm + 2 6 cm ____ cm	2 7 5 m + 3 3 1 m ____ m	2 4 5 km + 1 2 8 km ____ km

2. Subtract the following:

a	b	c
7 0 cm – 4 6 cm ____ cm	8 1 8 m – 3 4 9 m ____ m	9 3 2 km – 2 7 1 km ____ km

3. Multiply the following:

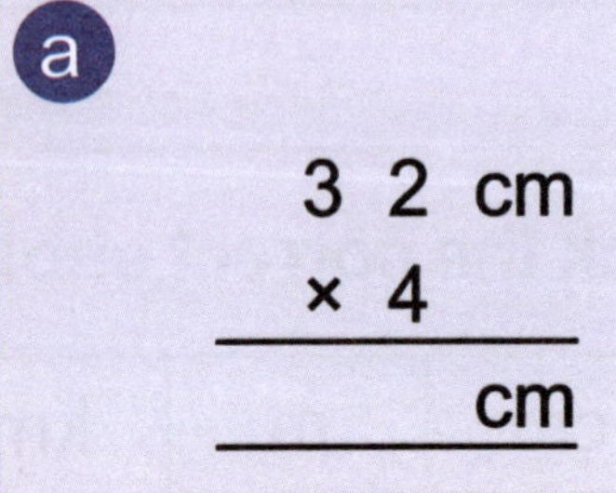

a	b	c
3 2 cm × 4 ____ cm	1 2 5 m × 6 ____ m	1 0 3 km × 8 ____ km

4. Divide the following:

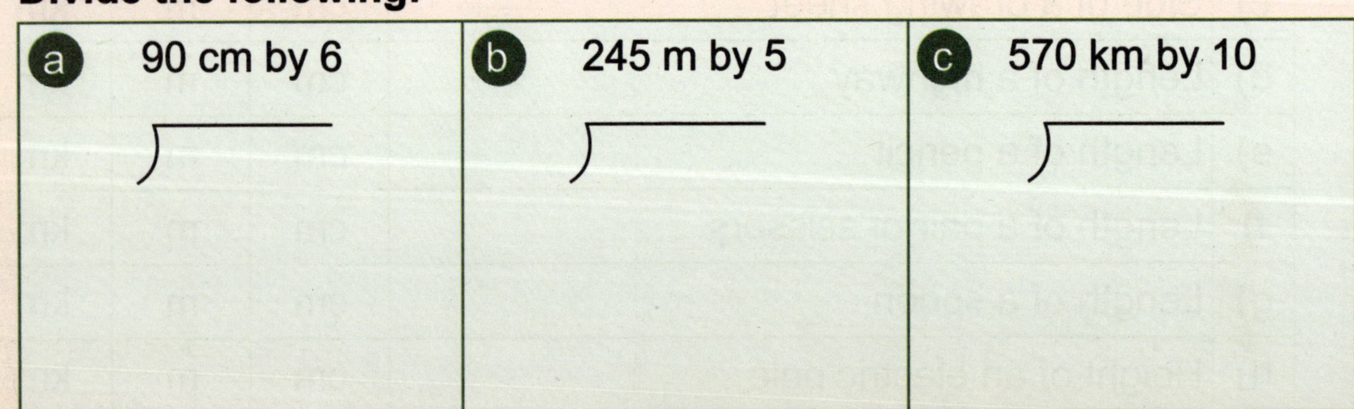

a	b	c
90 cm by 6	245 m by 5	570 km by 10

Worksheet - 23

Estimating Length

1. Round the following to nearest 10 cm.

a) 27 cm ≈ __________ cm   b) 42 cm ≈ __________ cm

c) A chord is 85 cm long. Its length rounded to nearest 10 cm is ______ cm.

2. Round the following to nearest 10 m.

a) 54 m ≈ __________ m   b) 328 m ≈ __________ m

c) A pathway is 237 m long. Its length rounded to nearest 10 m is _____ m.

3. Round the following to nearest 100 m.

a) 345 m ≈ __________ m   b) 673 m ≈ __________ m

c) A cable is 436 m long. Its length rounded to nearest 100 m is _____ m.

4. Round the following to nearest 10 km.

a) 52 km ≈ __________ km   b) 278 km ≈ __________ km

c) A highway is 123 km long. Its length rounded to nearest 10 km is _____ km.

5. Round the following to nearest 100 km.

a) 261 km ≈ __________ km   b) 856 km ≈ __________ km

c) Distance between two cities is 529 km. Distance rounded to nearest 100 km is __________ km.

6. Fill in the blanks with the correct sign (> or = or <).

a) 298 km rounded to nearest 10 km ______ 298 km rounded to nearest 100 km

b) 209 m rounded to nearest 10 m ______ 192 m rounded to nearest 10 m

c) 738 km rounded to nearest 10 km ______ 744 km rounded to nearest 10 km

d) 493 m rounded to nearest 10 m ______ 493 m rounded to nearest 100 m

e) 79 cm rounded to nearest 10 cm ______ 92 cm rounded to nearest 10 cm

f) 551 m rounded to nearest 100 m ______ 549 m rounded to nearest 100 m

g) 837 km rounded to nearest 100 km ______ 798 km rounded to nearest 100 km

Worksheet - 24

Length Stories

1. A path 437 m long was to be built. Of this 219 m has already been built. What length of path is left to be built?

2. City A is 243 km from city B, and city B is 218 km from city C. What distance do I have to travel if I go from A to B and then B to C?

3. A pipe is 12 m long. Another pipe is 15 m long. What is the total length of the two pipes?

4. Eight pieces of iron chains each measuring 25 m long were joined end to end to make a long chain. What is the length of this chain?

5. A cable 8 m long was cut into 8 equal parts. What is the length of each part in cm? [1 m = 100 cm]

Worksheet - 25

Choosing the Correct Unit

1. Fill in the blanks.

a) Units used to measure weight and their symbols are ________________ ____.

b) 1 kg = ____________ g

2. Which unit of weight will you use to measure the following? Tick the correct choice.

a)  g ☐ kg ☐	b)  g ☐ kg ☐	c)  g ☐ kg ☐
d)  g ☐ kg ☐	e)  g ☐ kg ☐	f)  g ☐ kg ☐
g)  g ☐ kg ☐	h)  g ☐ kg ☐	i)  g ☐ kg ☐
j)  g ☐ kg ☐	k)  g ☐ kg ☐	l)  g ☐ kg ☐

Worksheet - 26

Choosing the Correct Weights

Tick the weights required to weigh the following items:

	Item	Weight	Weights
1		650 g	500 g, 200 g, 100 g, 50 g
2		700 g	500 g, 500 g, 200 g, 50 g
3		12 kg	10 Kg, 5 Kg, 2 Kg, 1 Kg
4		7 kg	10 Kg, 5 Kg, 2 Kg
5		2 kg	5 Kg, 5 Kg, 2 Kg, 1 Kg
6		500 g	500 g, 200 g, 50 g, 50 g
7		16 kg	10 Kg, 5 Kg, 2 Kg, 1 Kg
8	 CEMENT	20 kg	10 Kg, 10 Kg, 2 Kg

Worksheet - 27

Addition, Subtraction, Multiplication, Division of Weights

1. Add the following:

a	b	c
6 5 g + 3 7 g ____ g	4 1 7 g + 3 8 9 g ____ g	2 3 8 kg + 2 7 5 kg ____ kg

2. Subtract the following:

a	b	c
6 5 g – 5 7 g ____ g	5 2 2 g – 1 6 5 g ____ g	7 9 5 kg – 4 5 9 kg ____ kg

3. Multiply the following:

a	b	c
1 7 g × 9 ____ g	2 8 g × 8 ____ g	1 1 7 kg × 6 ____ kg

4. Divide the following:

a	b	c
84 g by 6	235 g by 5	560 kg by 10

Worksheet - 28

Estimating Weight

1. Round the following to nearest 10 g:

a) 62 g ≈ ____________ g b) 457 g ≈ ____________ g

c) A mango weighs 315 g. Its weight rounded to nearest 10 g is ______ g.

2. Round the following to nearest 100 g:

a) 758 g ≈ ____________ g b) 732 g ≈ ____________ g

c) A pouch weighs 458 g. Its weight rounded to nearest 100 g is ______ g.

3. Round the following to nearest 10 kg:

a) 35 kg ≈ ____________ kg b) 542 kg ≈ ____________ kg

c) A carton weighs 349 kg. Its weight rounded to nearest 10 kg is _____ kg.

4. Round the following to nearest 100 kg:

a) 345 kg ≈ ____________ kg b) 653 kg ≈ ____________ kg

c) A motorcycle weighs 282 kg. Its weight rounded to nearest 100 kg is _____ kg.

5. Fill in the blanks with the correct sign (> or = or <).

a) 723 kg rounded to nearest 100 kg ______ 723 kg rounded to nearest 10 kg

b) 241 g rounded to nearest 10g ______185 g rounded to nearest 10g

c) 638 kg rounded to nearest 10 kg ______ 644 kg rounded to nearest 10 kg

d) 155 g rounded to nearest 10g ______ 155 g rounded to nearest 100g

e) 818 kg rounded to nearest 10 kg ______ 818 kg rounded to nearest 100 kg

f) 552 g rounded to nearest 100 g ______ 609 g rounded to nearest 100g

Worksheet - 29

Weight Stories

1. A pouch weighs 240 g. Another pouch weighs 475 g. What is the total weight of the two pouches?

2. A shop had 725 kg of wheat. Of this 429 kg was sold. How much wheat is left now?

3. A book weighs 250 g. How many kilograms will eight such books weigh?

4. 560 kg of salt is divided into seven equal parts. What is the weight of each part?

5. In one pan of a weighing balance we have a watermelon and a weight of 1 kg. In the other pan we have a weight of 5 kg. If the pans are balanced what is the weight of the watermelon?

Worksheet - 30

Measuring Capacity Using Non-Standard Units
Choosing the Correct Unit

1. Fill in the blanks.

a) The water in a [bucket] can fill 12 [mugs]. The capacity of the bucket is ________ mugs.

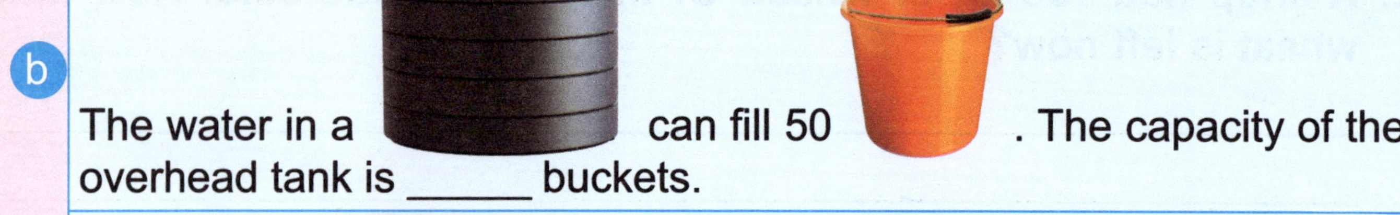

b) The water in a [tank] can fill 50 [buckets]. The capacity of the overhead tank is ______ buckets.

c) The water in a [jug] can fill 6 [bottles]. The capacity of the jug is ______ bottles.

d) The units used to measure capacity and their symbols are ___________

____.

e) 1 L = ______________ ml

2. What unit of capacity will you use to measure the following? Tick the correct choice.

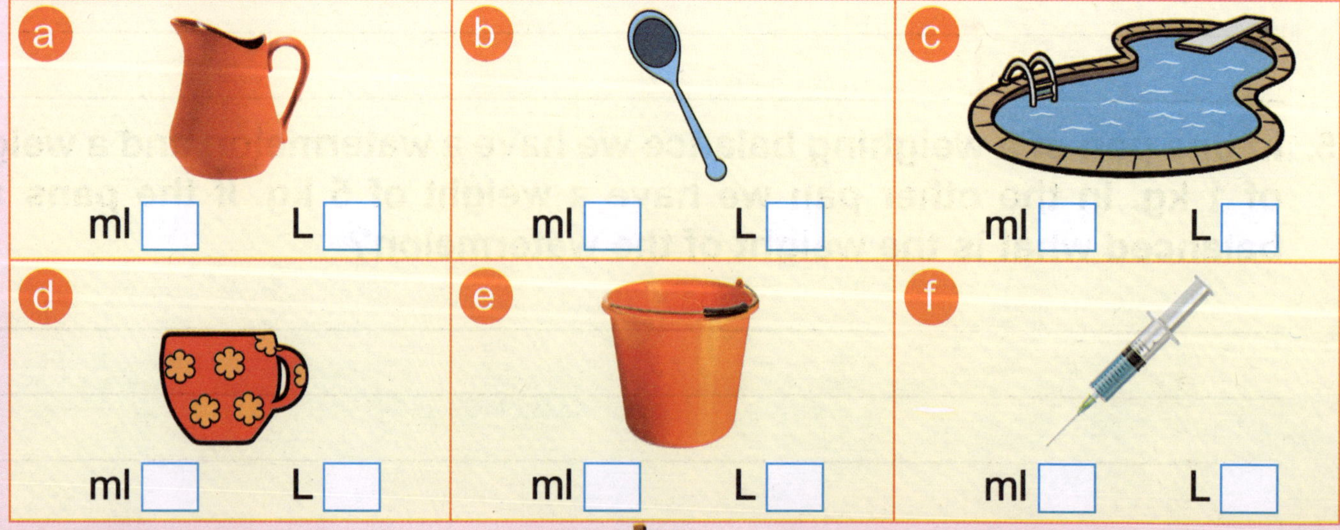

a) ml ☐ L ☐	b) ml ☐ L ☐	c) ml ☐ L ☐
d) ml ☐ L ☐	e) ml ☐ L ☐	f) ml ☐ L ☐

Worksheet - 31

Choosing the Correct Measuring Jug

Tick the measuring jug needed to measure the quantities given below.

1 250 ml	500 ml	200 ml	100 ml	50 ml
2 600 ml	500 ml	200 ml	100 ml	50 ml
3 850 ml	500 ml	200 ml	100 ml	50 ml
4 3 L	10 L	5L	2L	1L
5 350 ml	500 ml	200 ml	100 ml	50 ml
6 700 ml	500 ml	200 ml	100 ml	50 ml
7 7 L	10 L	5L	2L	1L
8 11 L	10 L	5L	2L	1L

Worksheet - 32

Addition, Subtraction, Multiplication, Division of Capacities

1. Add the following:

a	b	c
76 ml + 58 ml ____ ml	137 ml + 365 ml ____ ml	428 L + 377 L ____ L

2. Subtract the following:

a	b	c
58 ml – 29 ml ____ ml	719 ml – 258 ml ____ ml	738 L – 277 L ____ L

3. Multiply the following:

a	b	c
46 ml × 9 ml ____ ml	123 ml × 7 ml ____ ml	114 L × 8 L ____ L

4. Divide the following:

a	b	c
75 ml by 5	726 ml by 6	488 L by 8

Worksheet - 33

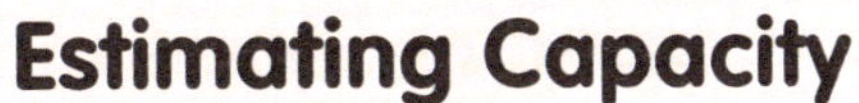

Estimating Capacity

1. Round the following to nearest 10 ml:

a) 75 ml ≈ ________________ ml   b) 53 ml ≈ ________________ ml

c) A bottle can hold 316 ml of milk when full. Its capacity rounded to nearest 10 ml is ________ ml.

2. Round the following to nearest 100 ml:

a) 567 ml ≈ ________________ ml   b) 875 ml ≈ ________________ ml

c) A tumbler can hold 469 ml of hair oil when full. Its capacity rounded to nearest 100 ml is ________ ml.

3. Round the following to nearest 10 L:

a) 67 L ≈ ________________ L   b) 642 L ≈ ________________ L

c) A drum can hold 358 L of oil when full. Its capacity rounded to nearest 10 L is ________ L.

4. Round the following to nearest 100 L:

a) 568 L ≈ ________________ L   b) 855 L ≈ ________________ L

c) An overhead tank can hold 911 L of water when full. Its capacity rounded to nearest 100 L is ________ L.

5. Fill in the blanks with the correct sign (> or = or <).

a) 825 L rounded to nearest 100 L _________ 825 L rounded to nearest 10 L.

b) 597 ml rounded to nearest 10 ml _________ 583 ml rounded to nearest 10 ml.

c) 438 L rounded to nearest 10 L _________ 444 L rounded to nearest 10 L.

d) 565 ml rounded to nearest 10 ml _________ 565 ml rounded to nearest 100 ml.

e) 698 L rounded to nearest 10 L _________ 688 L rounded to nearest 100 L.

f) 152 ml rounded to nearest 100 ml _________ 148 ml rounded to nearest 100 ml.

Worksheet - 34

Capacity Stories

1. A cup can hold 85 ml of tea. Another cup can hold 80 ml of tea. How much tea will the two cups have when full?

____

____

____

2. A milk tanker had 679 L of milk. It delivered 424 L at a milk booth. How much milk does it have now?

____

____

____

3. A bottle has 95 ml of oil. How much oil will 8 such bottles have?

____

____

____

4. 656 ml of sugar syrup was divided equally into 8 bowls. What is the quantity of the syrup in each bowl?

____

____

____

5. A tank had 587 L of water. Of this 312 L was used up, and 352 L of water was added to it. How much water does the tank hold now?

____

____

____

Days of the Week

Fill in the blanks.

1. The days of the week are Monday, ______________________________________

____.

2. ________________________ is the 4th day of the week.

3. ________________________ is the 6th day of the week.

4. ________________________ is the 1st day of the week.

5. Friday is the __________________ day of the week.

6. Wednesday is the ________________ day of the week.

7. Sunday is the ____________________ day of the week.

8. _________________________ comes before Friday.

9. _________________________ comes after Monday.

10. Saturday comes after _________________________.

11. Tuesday comes before _________________________.

12. _________________________ comes in between Friday and Sunday.

13. Tuesday comes in between _______________________ and Wednesday.

14. Sunday comes in between Saturday and _________________________.

Worksheet - 36

Months of the Year
Seasons of the Year

Fill in the blanks.

1. The months of the year are January, ______________________________

____

____.

2. March has ____________ days.
3. June has ____________ days.
4. October has ____________ days.
5. December has ____________ days.
6. February has ________ or ________ days.
7. ____________ comes before November.
8. April comes before ____________.
9. ____________ comes after October.
10. June comes after ____________.
11. July comes in between June and ____________.
12. October comes in between ____________ and November.
13. __________________ season comes between Summer and Autumn.
14. __________________ season comes after Spring.
15. __________________ season comes after Rainy season.
16. __________________ season comes before Spring.
17. __________________ season comes between Winter and Summer.

Worksheet - 37

Time on Clocks

Fill in the blanks.

1. A clock has __________ hands. A __________ hand and a __________ hand.
2. The ______________ hand tells us the minutes and ______________ hand tells us the hours.
3. The long hand of the clock is at 12 and short hand is at 2. What time is it? ______________

4	A train is to arrive at 3 o'clock. It arrives at the right time. Draw the hands on the clock to show the arrival time of the train.	
5	This clock is running two hours late. The correct time is _________________.	
6	This clock is running one hour fast. The correct time is _________________.	
7	This clock is running 2 hours late. The correct time is 11 o'clock. What time will this clock show? Draw hands. _________________	
8	This clock is running 3 hours fast. The correct time is 1 o'clock. What time will this clock show? Draw hands. _________________	

Worksheet - 38

Addition, Subtraction, Multiplication, Division of Money
Estimating Money

1. Add the following:

a	b	c
US$ 6 7	US$ 6 2 9	US$ 3 1 5
+ US$ 7 5	+ US$ 3 0 8	+ US$ 4 8 5
US$ ______	US$ ______	US$ ______

2. Subtract the following:

a	b	c
US$ 9 8	US$ 8 3 5	US$ 5 3 4
– US$ 6 4	– US$ 1 6 4	– US$ 3 6 7
US$ ______	US$ ______	US$ ______

3. Multiply the following:

a	b	c
US$ 4 5	US$ 1 2 6	US$ 1 0 2
× 6	× 7	× 8
US$ ______	US$ ______	US$ ______

4. Divide the following:

a	b	c
US$ 98 by 7	US$ 747 by 9	US$ 560 by 10

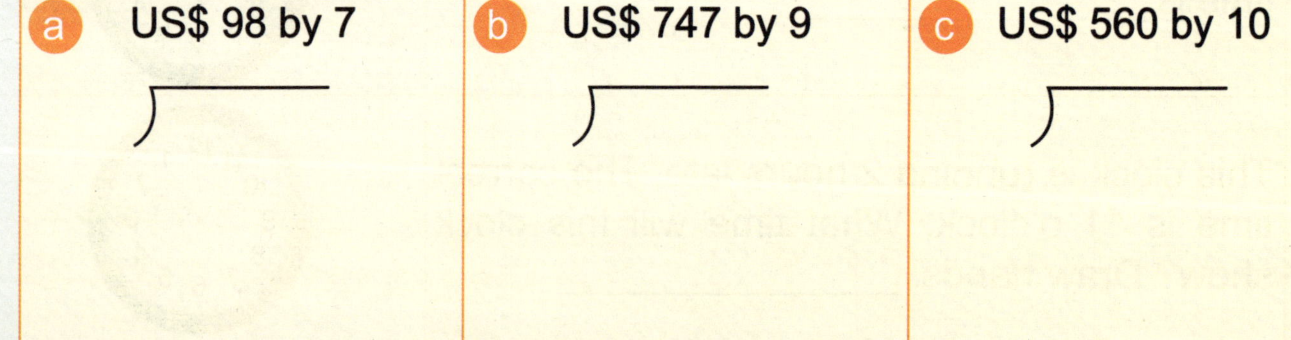

5. **A pair of trousers costs US$ 153. Its cost rounded to nearest 10 dollars is US$ ____________.**

6. **A bike costs US$ 739. Its cost rounded to nearest 100 dollars is US$ ____________.**

Worksheet - 39

Money Stories

1. Rosy bought a necklace for $ 656 and a ring for $ 175. How much money did she spend?

2. Nancy had nine notes of $ 100. She spent $ 625 on a laptop. How much money does she have now?

3. A shirt costs $ 145. How much will five such shirts cost?

4. Ten pairs of shoes of the same type cost $ 630. What is the cost of each pair of shoes?

5. Jasmine has three notes of $ 100, four notes of $ 50 and five notes of $ 10. How many dollars does she have in all?

Worksheet - 40

Patterns

1. What comes next?

2. Extend the sequence.

3. Fill in the blank boxes to complete the pattern.

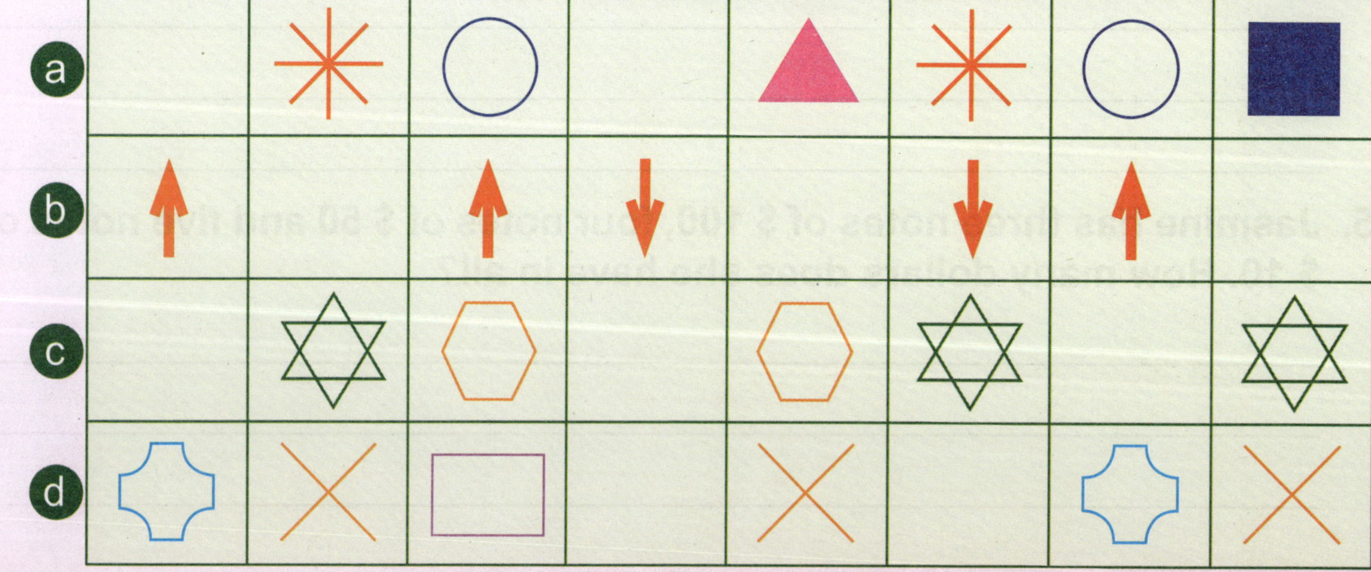

Worksheet - 41

Reducing and Growing Patterns

1. Fill in the empty boxes with the correct number.

a	112	223		445		667	778	889
b	123	234		456	567		789	890
c	AB-1		EF-3	GH-4		KL-6	MN-7	OP-8
d		B23	C34	D45	E56	F67		H89

2. Complete the following reducing patterns:

a	A B C D E F G H I	A B C D E F G H	A B C D E F G	A B C D E F			
b	1 2 3 4 5 6 7 8 9	1 2 3 4 5 6 7 8	1 2 3 4 5 6 7	1 2 3 4 5 6			
c	X X X X X X X X X	X X X X X X X X	X X X X X X X	X X X X X X			

3. Complete the following growing patterns:

a	0	0 0	0 0 0	0 0 0 0	0 0 0 0 0			
b	•	• •	• • •	• • • •	• • • • •			
c	☼	☼ ☼	☼ ☼ ☼	☼ ☼ ☼ ☼	☼ ☼ ☼ ☼ ☼			

Worksheet - 42

The Cricket Match

A cricket match was played between St. Stephen's School and St. George's School. It was a 20-over match. St. Stephen's School won the toss and decided to bowl first. Scoreboard after the match read as follows:

St. George's School		St. Stephen's School	
John	35	Tom	55
Peter	41	Jack	25*
David	12*	Fleming	27*
Alex	18*		
Total: 106 for 2 in 20 overs.		**Total: 107 for 1 in 19 overs.**	

(Note: * means not out)

Use the above information to fill in the blanks.

1. The cricket match was played between ____________________ and ____________________.
2. It was a _________ over match.
3. ____________________ won the toss.
4. ____________________ batted first. They batted for _________ overs and scored _________ runs losing _________ wickets.
5. _________ was the highest scorer for St. George's school.
6. ____________________ batted second. They batted for __________ overs and scored _________ runs losing _________ wicket.
7. _________ was the highest scorer for St. Stephen's school.
8. ____________________ won the match by _________ wickets.

Worksheet - 43

Horizontal, Vertical and Slanting Lines

Write H for horizontal, V for vertical and S for slanting in the boxes.

1 T

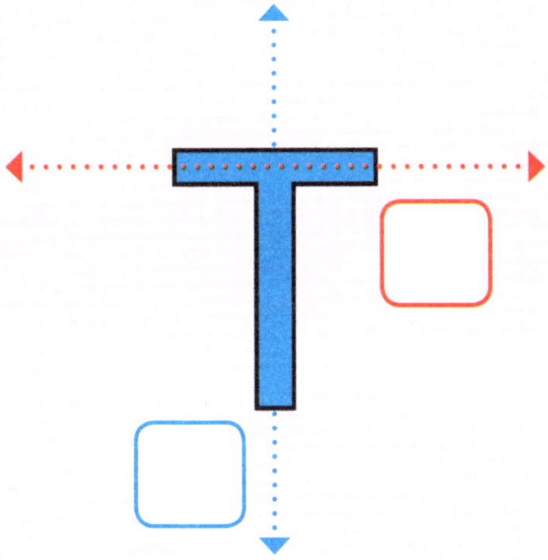

2 I

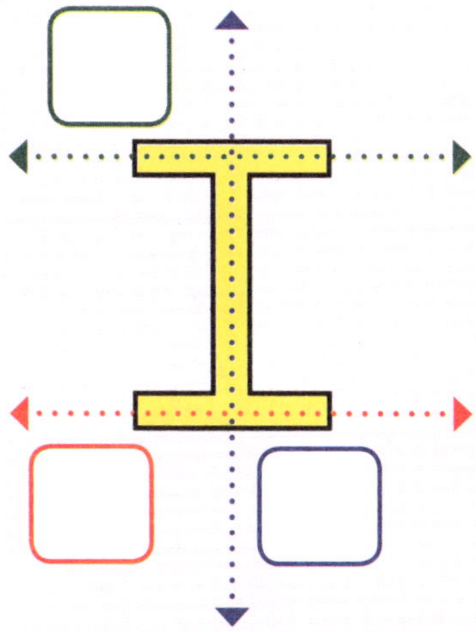

3 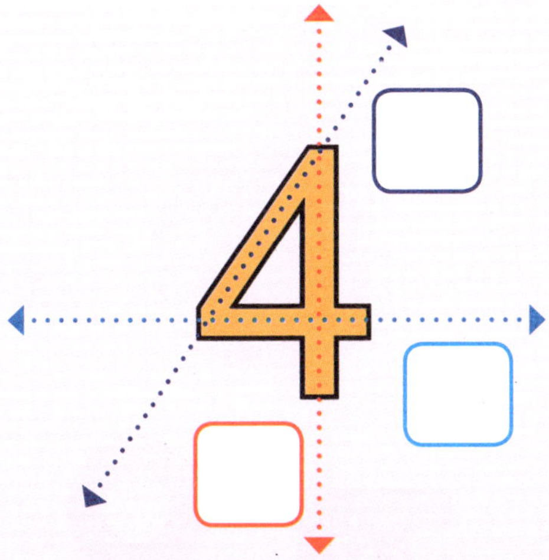 4

4 7

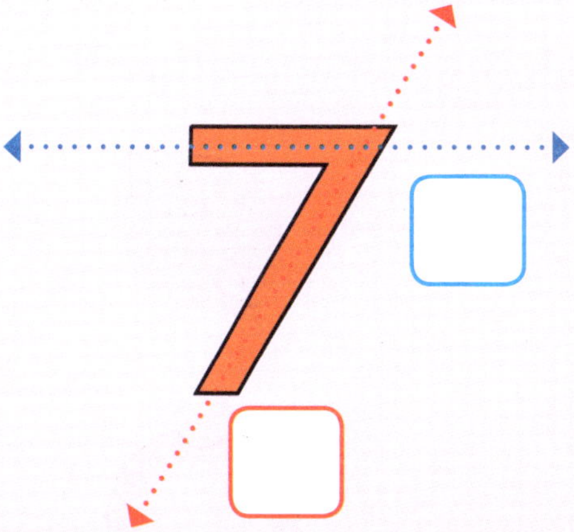

5

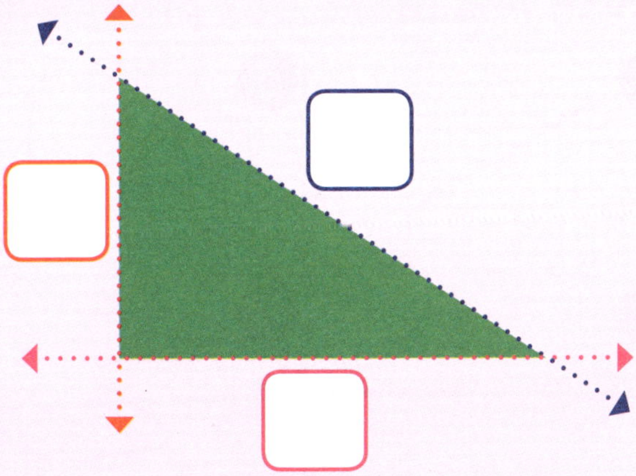

6

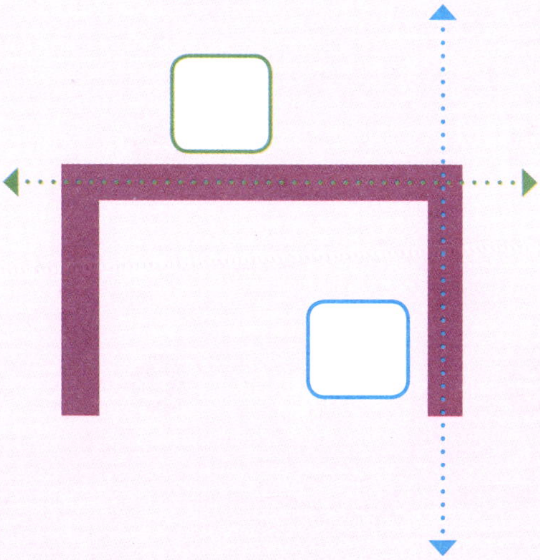

Worksheet - 44

Straight and Curved Edges

1. Fill in the blanks.

a		The scale has ______ edges. Each edge is a ______________________.
b		The hanger has ______ edges. Each edge is a ______________________.
c		The star has ______ edges. Each edge is a ______________________.
d		The coin has ______ edges. Each edge is a ______________________.
e		The envelope has ______ edges. Each edge is a ______________________.

2. Tick the curved edge for each of the following:

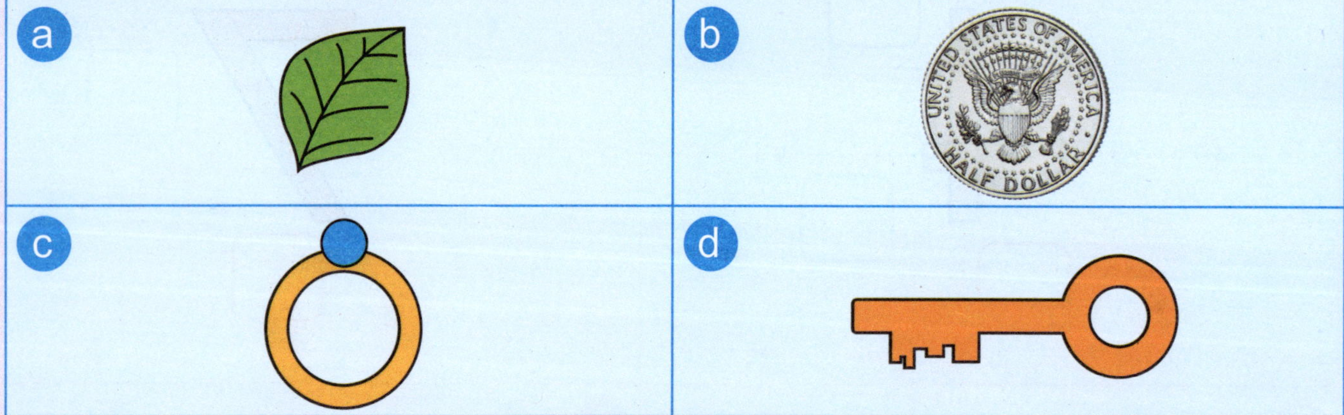

3. Write S for straight line and C for curved line in the box provided.

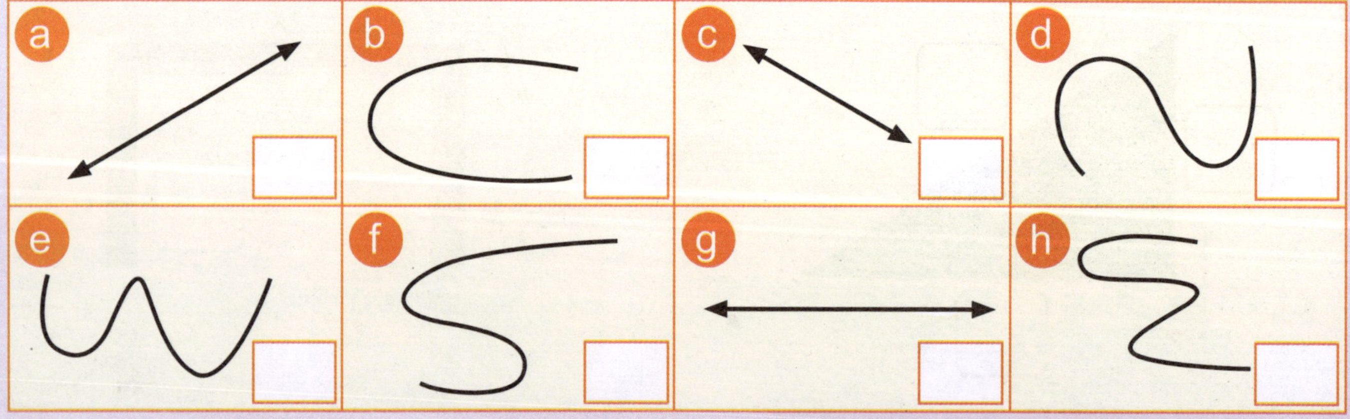

Worksheet - 45

Square and Rectangle

1. Tick the squares. Cross out the rectangles.

Number of squares ____________ Number of rectangles ____________

2. Fill in the blanks.

a) A square has _____ sides and _____ corners.

b) All sides of a square are __________. (equal/not equal)

c) A rectangle has _____ sides and _____ corners.

d) The two edges of a rectangle meeting at a corner are __________. (equal/not equal)

3. Tick the square objects. Cross out the rectangular objects.

Worksheet - 46

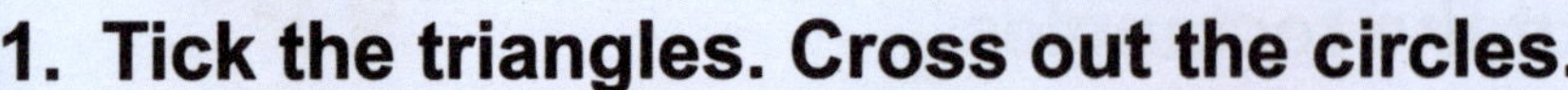

Triangle and Circle

1. Tick the triangles. Cross out the circles.

Number of circles ____________ Number of triangles ____________

2. Fill in the blanks.

a) A triangle has __________ sides and __________ corners.

b) A circle has a __________ edge.

c) A circle has __________ corners.

3. Tick the triangular objects. Cross out the circular objects.

Worksheet - 47

Matching Shadows

Match each picture with its correct shadow.

1 	a 
2 	b 
3 	c 
4 	d 
5 	e 

Worksheet - 48

Cube and Cuboid

1. Tick the cubes and cross out the cuboids.

Number of cubes ____________ Number of cuboids ____________

2. Fill in the blanks.

a) A cuboid has _____ faces, _____ edges and _____ corners.

b) The three edges of a cuboid meeting at a corner are ___________. (equal/not equal)

c) A cube has _____ faces, _____ edges and _____ corners.

d) All edges of a cube are _______________. (equal/not equal)

3. Tick the cube-shaped objects. Cross out the cuboid-shaped objects.

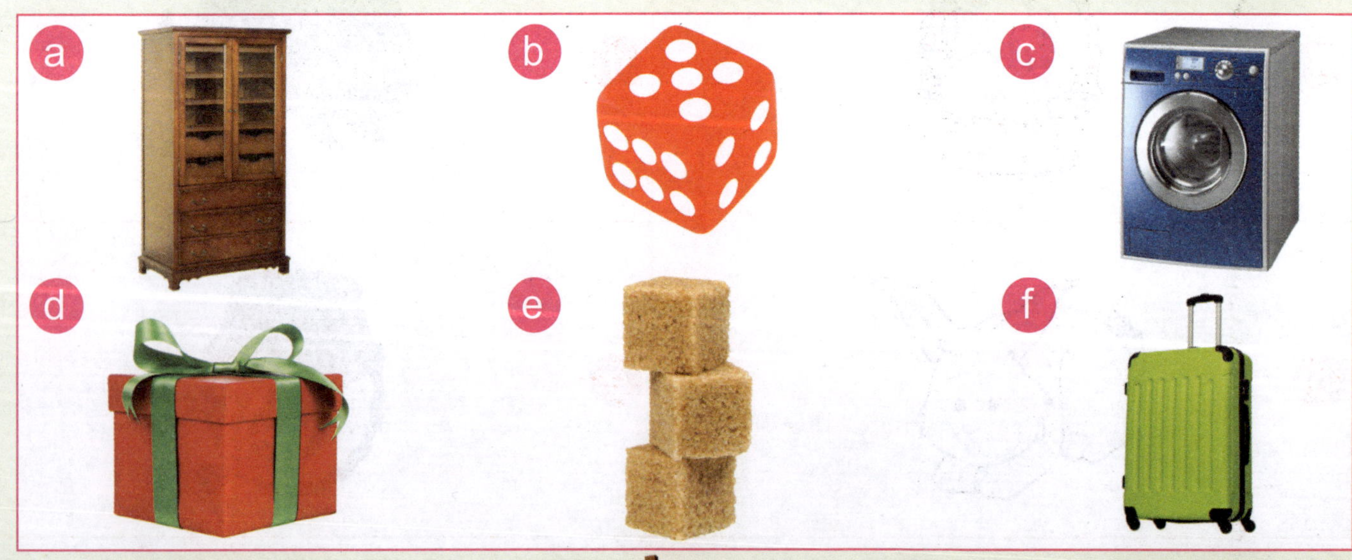

Worksheet - 49

Cone, Cylinder and Sphere

1. Tick the cones, circle the cylinders and cross out the spheres.

Number of cones ____ Number of cylinders ____ Number of spheres ____

2. Fill in the blanks.

a) A cone has _______ faces, _______ edge and _______ corner.

b) A cylinder has _______ faces, _______ edges and _______ corners.

c) A sphere has _______ face, _______ edges and _______ corners.

3. Circle the conical objects, cross out the cylindrical objects, and tick the spherical objects.

Worksheet - 50

2-D Outlines of 3-D Shapes

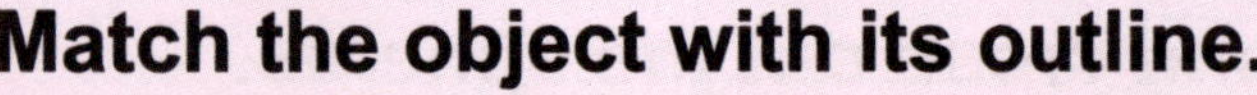

Match the object with its outline.

1	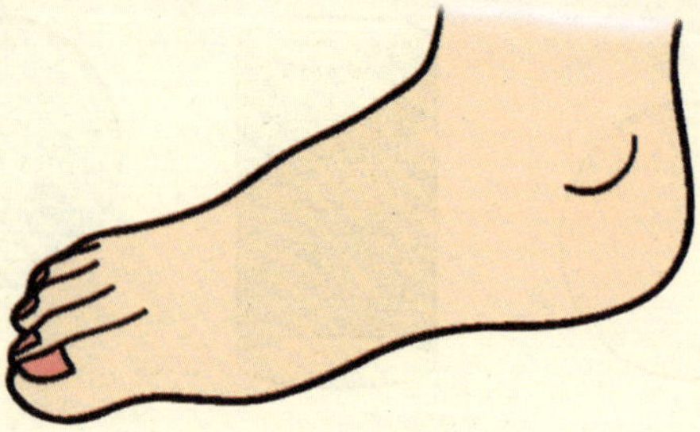 a 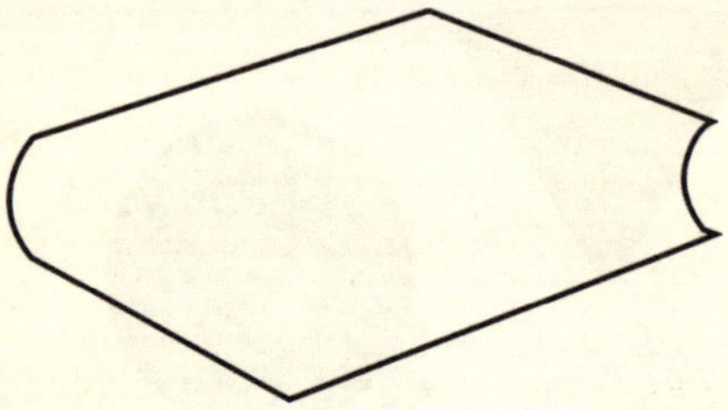
2	 b 
3	 c 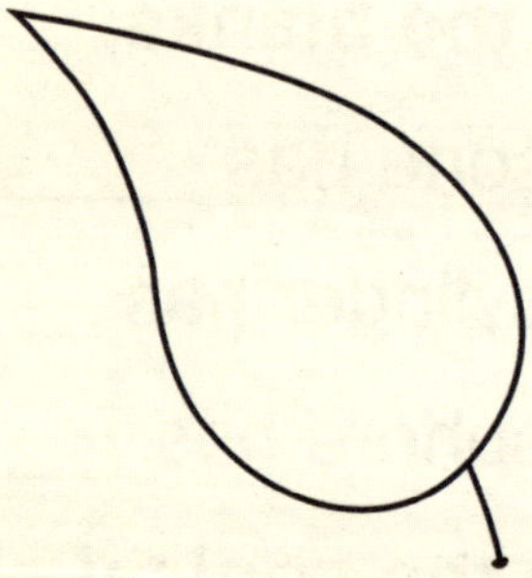
4	d 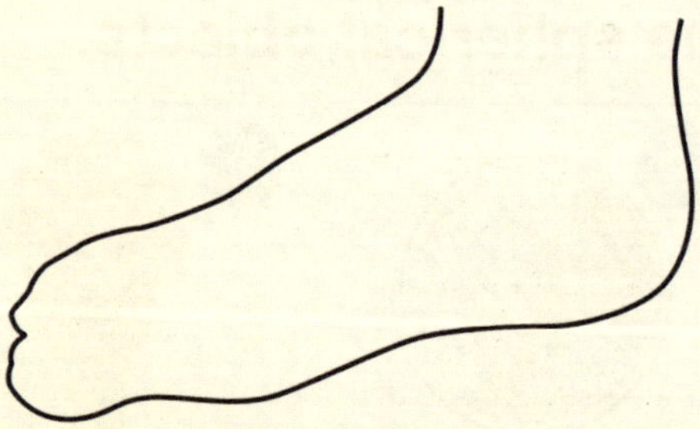
5	 e 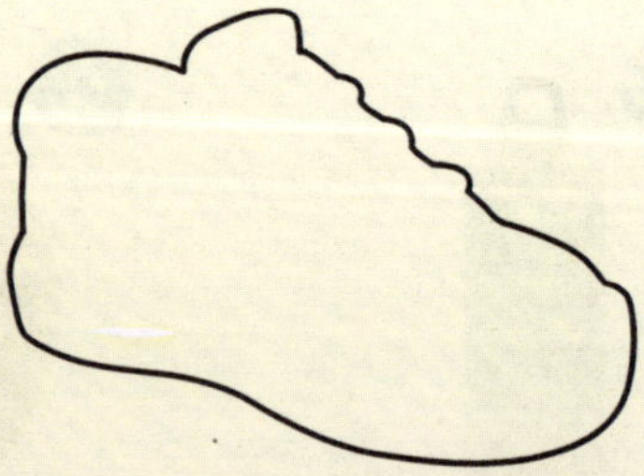

ANSWERS

Unit-1 Numbers: 100 to 1000

Worksheet-1

1 a)

113	114	115	116	117	118	119

b)

347	349	350	351	352

c)

553	554	556	557	558	559

d)

782	783	784	786

e)

992	993	994	995	996	998	999

2 a)

259	258	257	256	255	254

b)

432	431	429	428	427	426

c)

614	613	612	610

d)

701	700	699	698	697	696

e)

875	874	873	872	871

Worksheet-2

Numeral
126
245
369
434
496
512
563
651
680
740
780
843
890
930
992

Number name
One hundred eighty-five
Two hundred eighteen
Three hundred seventy-seven
Four hundred sixty-two
Four hundred eighty-nine
Five hundred twenty-nine
Five hundred eighty
Six hundred forty-three
Six hundred seventy-one
Seven hundred thirty-two
Seven hundred seventy-two
Eight hundred forty
Eight hundred seventy-five
Nine hundred twenty-nine
Nine hundred ninety-nine

Worksheet-3

1. a) 3; 4; 7 b) 4; 2; 8 2. a) 562 b) 659
3. a) Seven hundred seventy-four
   b) Eight hundred sixteen
4. a) 436 b) 654
   c) 950, 951 d) 379, 380, 381
5. a) 236 b) 532 c) 874, 875
   d) 463, 464, 465
6. a) 224 b) 326 c) 706, 707
   d) 997, 998, 999
7. a) 169, 171 b) 560, 561, 563, 564
   c) 994, 995, 996, 998, 999, 1000

Worksheet-4

1. a) 2; 5; 9 b) 6; 3; 8
2. a) 400; 10; 7 b) 700; 80; 0
3. a) 300; 60; 5 b) 900; 90; 8
4. a) 324 b) 518 c) 836
5. a) > b) = c) < d) = e) > f) <
6. a) 952, 534, 356, 348
   b) 834, 567, 449, 414, 352
7. a) 351, 354, 667, 753
   b) 314, 467, 500, 553, 922

Worksheet-5

1. 444; 200 2. 753; 135
3. 506, 560, 605, 650; G = 650, S = 506
4. a) 130 b) 230 c) 570 d) 760
5. 830 6. a) 300 b) 500 c) 900
   d) 1000 7. 600 8. >

Unit-2 Operations on Numbers

Worksheet-6

1. 817 2. 667 3. 794 4. 728
5. 916 6. 819 7. 558 8. 125
9. 421 10. 722 11. 461 12. 135
13. 377 14. 779 15. 755 16. 899
17. 385 18. 605 19. 621 20. 999

Worksheet-7

1. 677 2. 759 3. 570 4. 987
5. 108 6. 179 7. 209 8. 154
9. 461 10. 671 11. 558 12. 664
13. 617 14. 822 15. 633 16. 933
17. 438 18. 633 19. 680 20. 430

Worksheet-8

1. 370 2. 613 3. 623 4. 838 5. 501

Worksheet-9

1. a) 390 b) 780 c) 680 2. 300
3. a) 600 b) 900 c) 800 4. 800
5. a) 242 b) 423 c) 222 d) 526 e) 532
   f) 0 g) 614 h) 725 i) 829 j) 0

Worksheet-10

1. 283 2. 400 3. 364 4. 703
5. 628 6. 655 7. 454 8. 282
9. 388 10. 274 11. 202 12. 228
13. 530 14. 403 15. 241 16. 300
17. 424 18. 243 19. 438 20. 233

Worksheet-11

1. 481 2. 640 3. 394 4. 410
5. 655 6. 482 7. 691 8. 355
9. 350 10. 471

Worksheet-12

1. 463 2. 128 3. 51 4. 461 5. 605

Worksheet-13

1. a) 770 b) 340 c) 250 2. 420
3. a) 300 b) 400 c) 500 4. 100
5. a) 672 b) 739 c) 0 d) 756
e) 0 f) 533

Worksheet-14

6	7	8	9	10
12	14	16	18	20
18	21	24	27	30
24	28	32	36	40
30	35	40	45	50
36	42	48	54	60
42	49	56	63	70
48	56	64	72	80
54	63	72	81	90
60	70	80	90	100

1. 54 2. 28 3. 56 4. 45 5. 50

Worksheet-15

1. 606 2. 770 3. 888 4. 918 5. 858
6. 948 7. 875 8. 994 9. 936 10. 984
11. 945 12. 918 13. 660 14. 707 15. 800
16. 909 17. 942 18. 889 19. 952 20. 963
21. 760 22. 450 23. 720 24. 870

Worksheet-16

1. 852 2. 882 3. 920 4. 954 5. 960
6. 168 7. 162

Worksheet-17

1. a) B b) B c) 1; A d) 0; A
2. a) 209 b) 510 c) 512 d) 172
e) 623; 239 f) 882; 731 g) 563 h) 242
i) 372 j) 673 k) 526 l) 598
m) 261 n) 519 o) 1 p) 611
q) 872 r) 1 s) 0 t) 0
u) 0 v) 0 w) 0 x) 0
y) 0 z) 0

Worksheet-18

1. 61 2. 93 3. 117 4. 36
5. 82 6. 76; 2 7. 49; 4 8. 66; 7
9. 79; 4

Worksheet-19

1. a) 1 b) 598 c) 193 d) 1 e) 841 f) 1
2. a) 6; 10 b) 7; 5 c) 8; 5 d) 9; 4 e) 10; 6
3. a) 121 b) 73; 4

Unit-3 Measurement

Worksheet-20

1. 4 2. 5 3. 5 4. 7 5. 10 6. 8

Worksheet-21

1. a) 8 b) 4
2. a) 7 b) 8
3. a) metre (m), kilometre (km)
b) 100 c) 1000
4. a) km b) m c) cm d) km
e) cm f) cm g) cm h) m

Worksheet-22

1. a) 68 b) 606 c) 373
2. a) 24 b) 469 c) 661
3. a) 128 b) 750 c) 824
4. a) 15 cm b) 49 m c) 57 km

Worksheet-23

1. a) 30 b) 40 c) 90
2. a) 50 b) 330 c) 240
3. a) 300 b) 700 c) 400
4. a) 50 b) 280 c) 120
5. a) 300 b) 900 c) 500
6. a) = b) > c) = d) <
e) < f) > g) =

Worksheet-24

1. 218 m 2. 461 km 3. 27 m 4. 200 m
5. 100 cm

Worksheet-25

1. a) gram (g), kilogram (kg) b) 1000
2. a) g b) kg c) g d) g
   e) kg f) kg g) g h) kg
   i) g j) kg k) kg l) g

Worksheet-26

1. 500 g, 100 g, 50 g
2. 500 g, 200 g
3. 10 kg, 2 kg
4. 5 kg, 2 kg
5. 2 kg
6. 500 g
7. 10 kg, 5 kg, 1 kg
8. 10 kg, 10 kg

Worksheet-27

1. a) 102 b) 806 c) 513
2. a) 8 b) 357 c) 336
3. a) 153 b) 224 c) 702
4. a) 14 g b) 47 g c) 56 kg

Worksheet-28

1. a) 60 b) 460 c) 320
2. a) 800 b) 700 c) 500
3. a) 40 b) 540 c) 350
4. a) 300 b) 700 c) 300
5. a) < b) > c) =
   d) < e) > f) =

Worksheet-29

1. 715 g
2. 296 kg
3. 2 kg
4. 80 kg
5. 4 kg

Worksheet-30

1. a) 12 b) 50 c) 6
   d) millilitre (ml), litre (L) e) 1000
2. a) L b) ml c) L
   d) ml e) L f) ml

Worksheet-31

1. 200 ml, 50 ml
2. 500 ml, 100 ml
3. 500 ml, 200 ml, 100 ml, 50 ml
4. 2 L, 1 L
5. 200 ml, 100 ml, 50 ml
6. 500 ml, 200 ml
7. 5 L, 2 L
8. 10 L, 1 L

Worksheet-32

1. a) 134 b) 502 c) 805
2. a) 29 b) 461 c) 461
3. a) 414 b) 861 c) 912
4. a) 15 ml b) 121 ml c) 61 L

Worksheet-33

1. a) 80 b) 50 c) 320
2. a) 600 b) 900 c) 500
3. a) 70 b) 640 c) 360
4. a) 600 b) 900 c) 900
5. a) < b) > c) =
   d) < e) = f) >

Worksheet-34

1. 165 ml
2. 255 L
3. 760 ml
4. 82 ml
5. 627 L

Unit-4 Time and Money

Worksheet-35

1. Tuesday, Wednesday, Thursday, Friday, Saturday, Sunday
2. Thursday
3. Saturday
4. Monday
5. 5th
6. 3rd
7. 7th
8. Thursday
9. Tuesday
10. Friday
11. Wednesday
12. Saturday
13. Monday
14. Monday

Worksheet-36

1. February, March, April, May, June, July, August, September, October, November, December
2. 31
3. 30
4. 31
5. 31
6. 28 or 29
7. October
8. May
9. November
10. May
11. August
12. September
13. Rainy
14. Summer
15. Autumn
16. Winter
17. Spring

Worksheet-37

1. two; long; short
2. long; short
3. 2 o' clock
4.
5. 9 o' clock
6. 5 o' clock
7. 
   9 o' clock
8. 
   4 o' clock

Worksheet-38

1. a) 142 b) 937 c) 800
2. a) 34 b) 671 c) 167
3. a) 270 b) 882 c) 816
4. a) US$ 14 b) US$ 83 c) US$ 56
5. 150
6. 700

Worksheet-39

1. US$ 831 2. US$ 275 3. US$ 725
4. US$ 63 5. US$ 550

Unit-5 Patterns & Data Handling

Worksheet-40

1. a b c

2. a b c

3. a b c d

Worksheet-41

1. a) 334; 556 b) 345; 678 c) CD-2; IJ-5
d) A12; G78

2.

a	ABC DE	ABC D	ABC
b	123 45	123 4	123
c	XXX XX	XXX X	XXX

3.

a	000 000	000 000 0	000 000 00
b	••• •••	••• ••• •	••• ••• ••
c	☼☼☼ ☼☼☼	☼☼☼ ☼☼☼ ☼	☼☼☼ ☼☼☼ ☼☼

Worksheet-42

1. St. Stephen's School and St. George's School
2. 20 3. St. Stephen's School
4. St. George's School; 20; 106; 2
5. Peter
6. St. Stephen's School; 19; 107; 1
7. Tom 8. St. Stephen's School; 9

Unit-6 Geometry

Worksheet-43

1. V H 2. H H V
3. S H V 4. H S
5. S V H 6. H V

Worksheet-44

1. a) 4; straight line b) 3; straight line
c) 10; straight line d) 6; straight line
e) 4; straight line

2. a) b) c) d)

3. a) S b) C c) S d) C
e) C f) C g) S h) C

Worksheet-45

1. 12; 12
2. a) 4; 4 b) equal c) 4; 4 d) not equal
3. a) ✗ b) ✓ c) ✓ d) ✗
e) ✓ f) ✗ g) ✓ h) ✗

Worksheet-46

1. 10; 13
2. a) 3; 3 b) curved c) no
3. a) ✗ b) ✗ c) ✓ d) ✗
e) ✓ f) ✗ g) ✓ h) ✓

Worksheet-47

1 – c, 2 – d, 3 – a, 4 – e, 5 – b

Worksheet-48

1. 4; 4
2. a) 6, 12, 8 b) not equal c) 6, 12, 8
d) equal
3. a) ✗ b) ✓ c) ✗ d) ✓ e) ✓ f) ✗

Worksheet-49

1. 4; 4; 4
2. a) 2, 1, 1 b) 3, 2, no c) 1, no, no
3. a) ✓ b) O c) ✗ d) ✓ e) O
f) ✗ g) ✓ h) O i) ✗

Worksheet-50

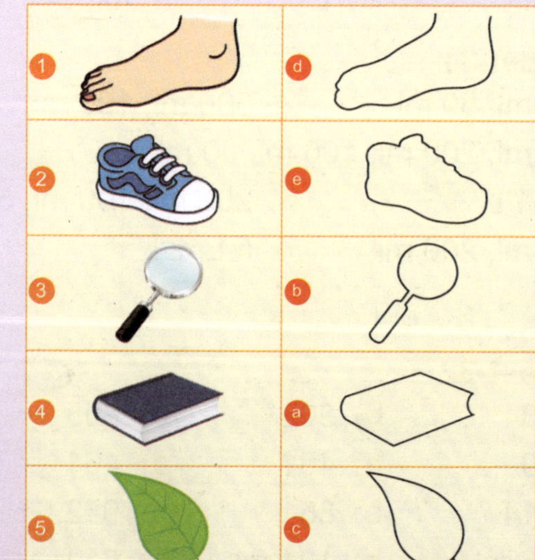

Printed in India